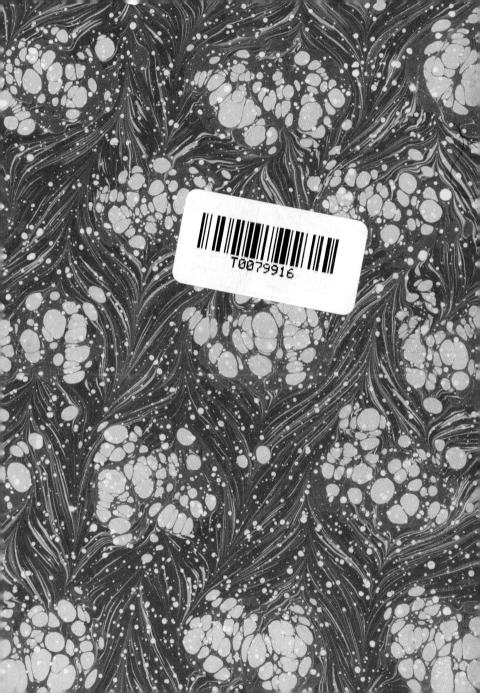

KEW POCKETBOOKS

FESTIVE FLORA

Curated by Mark Nesbitt and Lydia White

Kew Publishing
Royal Botanic Gardens, Kew

KEW HOLDS ONE OF THE LARGEST COLLECTIONS of botanical literature, art and archive material in the world. The Library comprises 185,000 monographs and rare books, around 150,000 pamphlets, 5,000 serial titles and 25,000 maps. The Archives contain vast collections relating to Kew's long history as a global centre of plant information and a nationally important botanic garden including 7 million letters, lists, field notebooks, diaries and manuscript pages.

The Illustrations Collection comprises 200,000 watercolours, oils, prints and drawings, assembled over the last 200 years, forming an exceptional visual record of plants and fungi. Works include those of the great masters of botanical illustration such as Ehret, Redouté and the Bauer brothers, Thomas Duncanson, George Bond and Walter Hood Fitch. Our special collections include historic and contemporary originals prepared for *Curtis's Botanical Magazine*, the work of Margaret Meen, Thomas Baines, Margaret Mee, Joseph Hooker's Indian sketches, Edouard Morren's bromeliad paintings, 'Company School' works commissioned from Indian artists by Roxburgh, Wallich, Royle and others, and the Marianne North Collection, housed in the gallery named after her in Kew Gardens.

INTRODUCTION

IT MIGHT SEEM TO BE A TRUISM THAT 'ALL LIFE depends on plants' – but it is remarkable how often plants are overlooked by humans. This has led botanists to adopt the language of 'ecosystem services' or 'nature-based solutions', rightly emphasising the practical benefits of plants to humankind. Many of these are long-known, for instance their value as food, medicine, clothing, fuel, and ornaments. To these we can now add carbon capture and cooling city streets, among many others. Naturally enough, research by ethnobotanists, who work at the intersection of people and plants, also focuses on the urgent problems of reconciling plant conservation and sustainable livelihoods.

Recently, another form of 'use' has attracted increasing interest: how people interact with plants in their daily lives, whether through the social aspects that surround the eating of a meal, or through caring for house plants or a garden, or through encounters with landscapes in the park or countryside. The pandemic lockdowns of 2020 have accelerated our understanding that many of the beneficial effects of plants on human wellbeing are intangible. They are therefore harder to measure – but of vital importance to mental wellbeing, community cohesion, and other markers of a healthy society.

The festive plants chosen for this book are reminders of just how deeply plant symbolism and rituals are embedded in human societies worldwide. Sometimes these uses are linked to religious ceremonies, but we have ranged broadly to include events that are culturally significant, or even linked to important local industries. Naturally, the European and North American Christmas features strongly, given its place as something of a global festival, but with help from Kew colleagues we have looked to other religions, places and times of year.

We have also taken a broad view of 'festive'; events such as Mexico's Day of the Dead, or Palm Sunday and Easter, or Sukkot are also solemn times of remembrance and reflection, nonetheless they involve joyful gatherings of family and friends. Some plant histories are solemn too. The cassava that is the essential ingredient in cassava pie was doubtless brought to Bermuda by ships engaged in the Atlantic slave trade.

The compelling stories associated with plant symbolism are a great way to engage people with plants, as can be seen in the recent flood of popular books on plant lore. This reveals the importance of the past: our encounters with plants are shaped by the stories told by our families, peers and through films, books and other media. Even where plant lore has been lost, it can be recovered from historical sources and shared again. Rediscovering and resharing the symbolism of plants can do much to enrich our lives.

The list on page 93 offers some signposts to further reading. As with other areas of plant history, there is much work to be done on unpicking plant symbolism. The subject has attracted rich accretions of its own mythology, and urgently needs more research based on careful analysis of original sources. Determining how long a plant has been used symbolically is tricky. For a start, it is surprisingly hard to translate many plant names in ancient texts, not least as the name has often been transferred to a more recent arrival.

Some continuities in meaning are highly plausible: it seems likely that laurel and olive have forever been culturally important in the eastern Mediterranean region. Other continuities might be true in general terms, for example the coincidence in month (though not in day) between the Christian Christmas and earlier markings of the shortest day of the year, such as Saturnalia. However, this does not mean that

today's Christmas tree (first seen in sixteenth-century Germany) has any connection to pre-Christian sacred trees. Similarly, the eighteenth-century English custom of kissing beneath the mistletoe has no connection to Pliny's remarks on its use by Druids in a very different context two millennia ago. However, its habit of remaining green and bearing abundant fruit in the depths of winter would plausibly explain its use as a winter decoration – and perhaps a connection to fertility.

This is also a rich field for contemporary observations: how are plants used in your community? What proportion of houses still have natural Christmas trees? What plants grow near places of worship or are laid on graves? What stories do children tell about local plants? There is plenty of research to be done. As the reader participates in the varied social and religious events that make up today's festivities, we encourage you to look out for plants and their stories.

Mark Nesbitt
Curator, Economic Botany Collection
Royal Botanic Gardens, Kew

Punica granatum

pomegranate, granada
from F. E. Köhler *Köhler's
Medizinal-Pflanzen*, 1887

———————

First taken into cultivation in or near to Turkey,
the blood-red juice and abundant seeds mean that
the fruit has often symbolised fertility, dawn,
life and death. The pomegranate tree has strong
association with the Zoroastrian communities of
Iran, being planted in temple courtyards. The
fruits are eaten by many Iranian families at Yaldā,
the festival of the longest night of the year.

Boswellia sacra

frankincense

from F. E. Köhler *Köhler's Medizinal-Pflanzen*, 1887

According to the Bible, three magi, or wise men from the east, brought frankincense, myrrh and gold as gifts to the infant Jesus. Frankincense and myrrh are the resin of wild trees growing in the Horn of Africa, where their harvesting supports local livelihoods. Both resins are still an important ingredient in church incense.

Tagetes erecta

marigold, flor de muerto, African marigold
by D. Bois from Edward Step *Favourite Flowers
of Garden and Greenhouse*, 1896

Remarkably, the yellow-orange marigold flower
is an important part of religious ritual both in its
native home of Mexico, and in India. In Mexico,
the cempasúchil, sacred since Aztec times, is used
to decorate the graves for the Day of the Dead.
In India, marigold garlands – considered
auspicious – are widely prevalent in religious
festivals and public ceremonies.

Manihot esculenta

cassava, tapioca, manioc, yuca
from Michel Étienne Descourtilz
Flore Pittoresque et Médicale des Antilles, 1821–9

Manioc tubers are a staple food in the Amazon
rainforest, and have spread from this region to
throughout the wet tropics. Cassava pie is a
ubiquitous tradition on the island of Bermuda,
eaten in every household on Christmas day.
It is made of grated cassava, mixed to make a
sweet dense batter and baked in two layers
with chicken or pork in-between.

Rosa × damascena

rose
by Pierre Joseph Redouté,
Kew Collection, 19th century

Most of the world's production of attar of rose
takes place in central Bulgaria. Since 1903 the
town of Kazanlak has hosted a rose festival every
June. Ceremonies celebrate the harvest of the
rose petals from which the essential oils are
extracted. The oil is widely used in perfumery,
and rose water in the preparation of sweets for
festive occasions throughout the Middle East.

Caryota urens

toddy palm, East Indian wine palm, fishtail palm
from Michel Étienne Descourtilz
Flore Pittoresque et Médicale des Antilles, 1821–9

The toddy palm is native to India and Sri Lanka.
Its flowers are tapped for a sweet sap that can be
boiled down in an open pan to make jaggery.
This can be used as a sweetener, or fermented
to a palm wine (toddy). At the Hindu festival of
Sankashti Chaturthi, in honour of Ganesha, the
god of wisdom, sweets made of toddy palm
and sugarcane jaggery are offered to the
god and consumed by his devotees.

Abies nordmanniana

Christmas tree, Nordmann fir
from Cornelis Antoon Jan Abraham Oudemans
Neerland's Plantentuin, 1865

———————

The traditional Christmas tree was introduced to
Britain in 1800 by George III's German wife,
Queen Charlotte. She was a keen botanist, often
resident in the royal palace at Kew Gardens.
About 8 million Christmas trees are now sold
each year in the UK. The Nordmann fir is native
to Turkey and the Caucasus, but widely grown for
its compact form and leaves that don't drop.

Galium verum

lady's bedstraw, drăgaica, sînziene
from William Curtis *Flora Londinensis*, 1775–98

———————

Midsummer Day (Sanzienele) is still celebrated in
rural Romania, with pre-Christian elements of the
summer solstice celebrating fertility combined
with a Christian festival marking the birth of
St. John the Baptist. The flowers are picked by
young women and consecrated at the church.
They are then placed at the gate of the household
to protect its members from evil.

Plumeria alba

frangipani
from Nicolaus Joseph von Jacquin
Selectarum Stirpium Americanarum Historia,
1780–1

The frangipani tree has large, fragrant blossoms,
used for making lei in Hawaii. On White Sunday
(Lotu Tamaiti) in Samoa, children attend
church dressed in white and wearing a crown
of white frangipani blossoms.

Ocimum tenuiflorum

holy basil, tulsi, sacred basil
by Marianne North from the
Marianne North Collection, Kew, 1878

Closely related to sweet basil, holy basil is an
important sacred plant in Hinduism. Many
Hindu homes and temple courtyards contain
a plant. It is believed to protect against
misfortune and represents purity,
harmony, serenity and luck.

Euphorbia pulcherrima

poinsettia
from Benjamin Maund and
John Stevens Henslow *The Botanist*, 1838

This beautiful species of spurge grows wild in
the forests of Mexico and Guatemala. Before the
Spanish conquest, it was known to the Aztecs as
cuitlaxochitl, and imported in large quantities to
their capital (now Mexico City) for religious
purposes. Worldwide, the potted plants are today
popular at Christmas for the contrast between
their green foliage and the scarlet red bracts
that surround the flowers.

Curcuma longa

turmeric
by Elizabeth Blackwell from Elizabeth Blackwell
Herbarium Blackwellianum, 1750–73

Turmeric is a member of the ginger family
native to the Indian subcontinent. Its
underground rhizomes are used for dye and
as a spice and medicine, and it is widely used
in all culinary traditions across India.
Yellow and orange are colours with spiritual
connotations in the Indian subcontinent.

Vitis vinifera

grape
from Antonio Targioni Tozzetti
Raccolta di Fiori Frutti ed Agrumi, 1825

The grape harvest is celebrated in winemaking areas of Hungary by a harvest festival or szüreti bál. Dance rooms are decorated with bunches of grapes, to be sold to the highest bidder.

N° 14

Cinnamomum verum

cinnamon
from F. E. Köhler *Köhler's
Medizinal-Pflanzen*, 1887

Cinnamon is one of several tropical spices, such
as nutmeg and mace, strongly associated with the
European celebration of Christmas. It grows wild
in Sri Lanka, where it is now also a major crop.
As well as use in cakes and puddings, in parts of
Germany it is used to make New Year's Eve
(Silvesterabend) punch, with red wine and sugar.

Phoenix dactylifera

date palm
from Henri Louis Duhamel du Monceau
*Traité des Arbres et Arbustes que l'on Cultive en
France en Pleine Terre*, 1800–19

———————

The date palm is grown in warmer parts of north
Africa, the Middle East, and South Asia, for its
sweet fruit that are staple in local diet. The tree
is considered a blessing and every part of the
plant is used in the Arab world. Palm Sunday,
in Christianity, marks the day of the Jewish
Passover on which Jesus is described as riding
into Jerusalem to be greeted by waving palm
branches, most likely of date palm.

Amaranthus cruentus

amaranth

by P. J. F. Turpin from Jean Louis Marie Poiret
Leçons de Flore, 1819

Amaranth is not a grass, but its starchy seeds
have similar properties to wheat or maize.
Amaranths are ancient grains of Mesoamerica
and the Andes, and their popped grains have long
been used as food and in Aztec religious rituals.
Today, skulls known as 'calaveritas de amaranto'
are made from the popped seeds, held together
by raw sugar, and consumed during Mexico's
Day of the Dead holiday.

Dioscorea alata

yam, purple yam, white yam
from Michel Étienne Descourtilz
Flore Pittoresque et Médicale des Antilles, 1821–9

———————

The New Yam festival is widely celebrated in
Nigeria. Held at the end of June, it marks the
moment that the crop can be harvested and eaten.
The custom spread with enslaved Africans to Haiti,
where the festival of Manger Yam likewise
celebrates the new harvest.

Malus domestica

apple

from Pierre Antoine Poiteau
Pomologie Française, 1846

Apples were first grown in the Tian Shan
mountains of Central Asia, crossing with local
crab apples as they spread to Europe. The
Annapolis Valley in Nova Scotia, Canada, hosts
an Apple Blossom Festival each year. In Britain
and France the apple harvest season is often
the occasion for local cider festivals.

Lawsonia inermis

henna

from William Roxburgh *Flora Indica*, 1820–4

Henna originated in the Near East, but its use
in decorating hands and feet as temporary art
has spread to India and other regions of Asia and
Africa. The reddish-brown dye is made from the
ground leaves, and is used as part of many
ceremonies, including Hindu weddings and the
Islamic festivals of Eid al-Fitr and Eid al-Adha
that mark the end of Ramadan and Hajj. Henna
symbolises prosperity, fertility, happiness,
fortune, seduction and beauty.

Crataegus monogyna

hawthorn

from Georg Christian Oeder *Flora Danica*, 1761

There is a rich vein of folklore in the British countryside associated with the hawthorn, which is considered both lucky and unlucky, often linked to the month of May. In the village of Appleton Thorn, Cheshire, children still dance around the hawthorn tree on Bawming the Thorn day.

Zingiber officinale

ginger

from F. E. Köhler *Köhler's Medizinal-Pflanzen*, 1887

Ginger comes from the rhizomes (underground stem) of this tropical plant. Dried ginger has been exported from India to Europe since Roman times. It is a common ingredient in the Lebkuchen, or honey cakes, made in many German cities and much consumed at Christmas. Highly decorated gingerbread is a feature of harvest celebrations in Transylvania, Romania.

Trifolium repens

white clover, shamrock, seamróg
from Otto Wilhelm Thomé *Flora von
Deutschland Österreich und der Schweiz*, 1885

––––––––––––

White clover is one of several plants commonly
identified as shamrock. It has a long association
with St. Patrick, the patron saint of Ireland,
perhaps because the three leaves of a clover sprig
were seen to represent the Christian Trinity.
Since the eighteenth century, shamrock has
become the national symbol of Ireland.

Zea mays

corn, maize
from *Flore des serres et des jardin de l'Europe*, 1873

Since it was first taken into cultivation in Mexico more than 7,000 years ago, maize has spread worldwide as a grain and as animal feed. It is an important symbol in Kwanzaa, a winter celebration of African-American culture held since 1966. An ear of corn is set out for each child in the family.

Betula pubescens

birch, **downy birch**
from Georg Christian Oeder *Flora Danica*, 1761

In European folklore the boughs of birch have
long been associated with protective powers
against evil spirits. In Denmark and Norway
birch twigs feature at the festival of Fastelavn,
in the days before Lent. Children decorate
bundles of rods – these days from varied
trees – and decorate them with sweets.

Theobroma cacao

cacao, cocoa tree
from F. E. Köhler *Köhler's*
Medizinal-Pflanzen, 1887

Today chocolate, made from the seeds of the
cacao tree, is often found in festive gifts such as
Easter eggs. Cacao was an essential component of
religion and ceremony in the ancient societies
of Mesoamerica, consumed during weddings,
offered as foaming chocolate drinks to the dead,
and widely used in food and medicine.

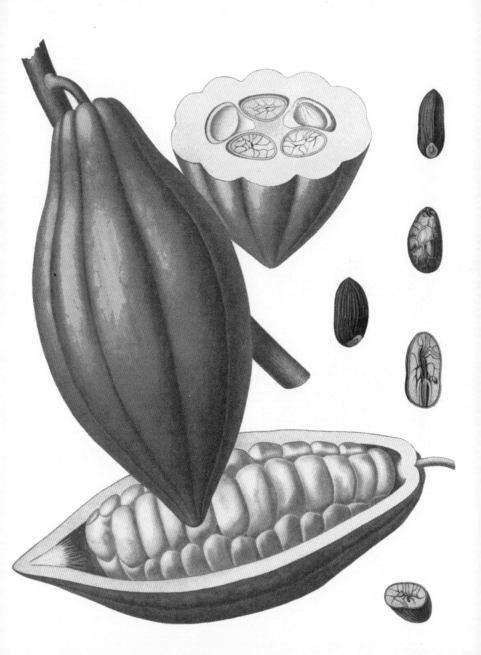

88

Olea europaea

olive

from F. E. Köhler *Köhler's
Medizinal-Pflanzen*, 1887

The olive tree is integral to Mediterranean life,
both as an element of landscape, and through the
myriad products of its oil. Olive is often mentioned
in the Bible, and it retains symbolic importance
in all those religions that originate in the Near
East. In Christianity olive branches are often
associated with Palm Sunday.

Aframomum melegueta

grains of paradise
by W. H. Fitch from
Curtis's Botanical Magazine, 1801

Grains of paradise are the seeds of a member of
the ginger family, native to west Africa. They
have a pungent, peppery flavour and are much
used in the cuisine of west and north Africa,
and formerly as a flavouring for beer in England.
The spice is important in west African culture
and medicine, for example as a gift at Yoruba
births and weddings.

Ilex aquifolium

holly

from Thomas Green *The Universal Herbal*, 1816

The evergreen holly with its red berries is a key
element of the Christmas wreath, hung on the
front door. This custom may hark back to the
Roman festival of Saturnalia, held at the same
time of year, during which wreaths of holly
and ivy were similarly hung.

Prunus species

cherry

from Yokohama Ueki Kabushiki Kaisha
Catalogue of the Yokohama Nursery Co., Ltd, 1907

The popular custom of hanami (flower viewing)
in Japan takes place in March and April, often
accompanied by picnicking under the trees.
Gifts of cherry trees from Japan have spread
this springtime custom to the shores of the
Tidal Basin in Washington, DC.

Salix babylonica

willow

by Pierre Joseph Redouté from Henri Louis
Duhamel du Monceau *Traité des Arbres et
Arbustes que l'on Cultive en France en
Pleine Terre*, 1800–19

The fuzzy buds of willows are a very visible sign of
spring. It's not surprising that they are often seen
at the Christian festival of Easter, particularly in
the Orthodox churches, where they are strongly
associated with the rising of the dead by Lazarus.

Nelumbo nucifera

lotus, Indian lotus, sacred lotus
by Marianne North from the
Marianne North Collection, Kew, 1876

The lotus flower is an important symbol in
Buddhism, Hinduism, and Confucian culture.
It stands for eternity, plenty, and abundant good
fortune. The flowers are used as offerings in
Buddhist and Hindu temples.

Laurus nobilis

bay tree, laurel
by Elizabeth Blackwell from Elizabeth Blackwell
A Curious Herbal, 1737

—————————

In ancient Greece laurel was the symbol of the
god Apollo. Today, alongside its role as a culinary
herb, laurel is a symbol of the resurrection in
Christianity, for example in the Greek Orthodox
Easter. Laurel wreaths also feature as winners'
crowns, as at the ancient Pythian Games, and
in ceremonies of remembrance.

Cocos nucifera

coconut

from F. E. Köhler *Köhler's
Medizinal-Pflanzen*, 1887

The coconut palm is a hugely important source
of food and material throughout the tropics.
It is used as an offering, and in many celebratory
foods. In Sri Lanka coconut milk is used in
savoury and sweet dishes, while in South-East
Asia strips of coconut leaf are used to wrap
ketupat, a rice cake that is eaten at the end of
the Ramadan fasting period.

Arctium lappa

greater burdock
from Jan Kops *Flora Batava*, 1800–46

Once a year since medieval times the Burryman
has walked the streets of South Queensferry,
southern Scotland. With his body thickly coated
in burdock heads, he walks from house to house,
receiving nips of whisky and bringing good luck.

Acacia pycnantha

golden wattle
from John Ednie Brown *The Forest Flora
of South Australia*, 1882–93

———————

The golden wattle is the national flower of
Australia. National Wattle Day has been
celebrated on 1 September since 1990, with a
history going back a hundred years before that.
Special events and the wearing of a sprig of
wattle mark a day that sets out to celebrate all
Australians, whether indigenous or settler.

Chrysanthemum species

chrysanthemum
by D. Bois from Edward Step *Favourite Flowers
of Garden and Greenhouse*, 1896

The Chrysanthemum Festival was one of the
five sacred festivals of Japan. Statues were
exhibited in September and October, created by
guiding chrysanthemum plants in wire frames so
that the flowers covered the surface. Exhibitions
of such kiku ningyo are now uncommon, but
chrysanthemum cakes and wine are occasionally
consumed to mark the festival.

Crocus sativus

saffron, saffron crocus
by Elizabeth Blackwell from
Elizabeth Blackwell *A Curious Herbal*, 1737

———

Saffron appears often in the extraordinary
3,600-year-old Minoan frescoes at Akrotiri,
on the Aegean island of Santorini. It clearly
had symbolic importance beyond its uses as a
dye, medicine and spice. Today most saffron
is grown in Iran. As an expensive spice, it is
used in foods to celebrate special occasions,
such as simnel cake at Easter, or saffron
buns for St Lucia's day in Sweden.

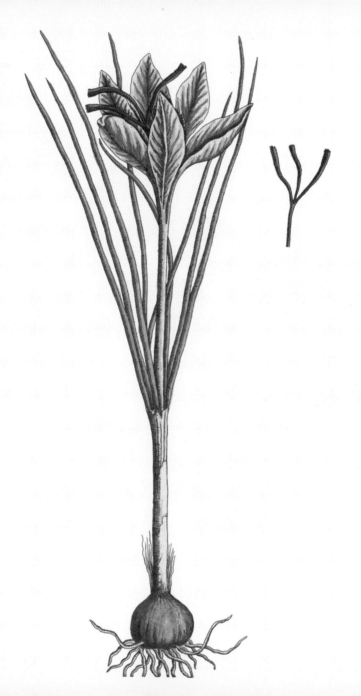

Citrus medica

citrus, citron, etrog
from Antonio Targioni Tozzetti
Raccolta di Fiori Frutti ed Agrumi, 1825

The citron tree was taken from the foothills
of the Himalayas to the Near East about
2,500 years ago, becoming widely grown in
Roman times. Etrog plays an important role
in the Jewish festival of Sukkot, marking the
harvest of the autumn fruits.

N⁰ 39

Viscum album

mistletoe
from F. E. Köhler *Köhler's
Medizinal-Pflanzen*, 1887

———————

Mistletoe grows throughout much of Europe,
as a parasite on trees such as apple and
hawthorn. The pearl-like berries ripen at
Christmas time, perhaps accounting for its
widespread association with fertility.

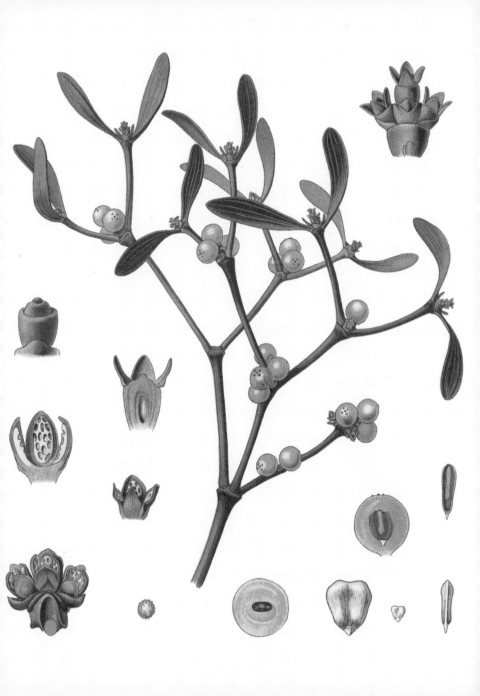

Begonia species

begonia
from *Revue Horticole*, 1890

Belgium grows over half the world's begonia
tubers, mostly for export. The beauty of this
ornamental flower has been celebrated for
80 years at the Lochristi Begonia Festival,
held every August near Ghent.

ILLUSTRATION SOURCES

Books and Journals

Blackwell, E. (1737). *A Curious Herbal*. Volume I. J. Nourse, London.

Blackwell, E. (1750–73). *Herbarium Blackwellianum*. Volume IV. Typis Io. Iosephi Freischmanni, Nuremberg.

Brown, J. E. (1882–93). *The Forest Flora of South Australia*. E. Spiller, Adelaide.

Carrière, E. A. (1890). Bégonias tubéreux multiflores. *Revue Horticole*. Librairie Agricole de la Maison Rustique, Paris.

Curtis, J. (1823). *British Entomology*. Volume VI. London.

Curtis, W. (1775–98). *Flora Londinensis*. Volume VI. London.

Descourtilz, M. É. (1821–9). *Flore Pittoresque et Médicale des Antilles*. Volume III, VII & VIII. Chez Corsnier, Paris.

Duhamel du Monceau, H. L. (1755). *Traité des Arbres et Arbustes qui se Cultivent en France en Pleine Terre*. Volume I. H. L. Guerin & L. F. Delatour, Paris.

Duhamel du Monceau, H. L. (1800–19). *Traité des Arbres et Arbustes que l'on Cultive en France en Pleine Terre*. Volume III & IV. Didot *et. al.*, Paris.

Green, T. (1816). *The Universal Herbal*. Volume I. Caxton Press, Liverpool.

Hooker, W. J. (1801). *Amomum Granum Paradisi. Curtis's Botanical Magazine*. Volume 77, t. 4603.

Jacquin, N. J. von (1780–1). *Selectarum Stirpium Americanarum Historia*. N. J. von Jacquin, Vindobona.

Köhler, F. E. (1887). *Köhler's Medizinal-Pflanzen*. Volume I, II & III. F. E. Köhler, Gera-Untermhaus.

Kops, J. (1800-46). *Flora Batava*. J. C. Sepp en Zoon, Amsterdam.

Maund, B. and Henslow, J. S. (1838). *The Botanist*. Volume II. R. Groombridge, London.

Oeder, G. C. (1761). *Flora Danica*. Copenhagen.

Oudemans, C. A. J. A. (1865). *Neerland's Plantentuin*. Gebr. van der Post, Utrecht.

Poiret, J. L. M. (1819). *Leçons de Flore*. Volume II. C. L. F. Pancoucke, Paris.

Poiteau, P. A. (1846). *Pomologie Française*. Volume IV. Langlois et Leclercq, Paris.

Roxburgh, W. (1820–4). *Flora Indica*. Volume II. Mission Press, Serampore.

Step, E. (1896). *Favourite Flowers of Garden and Greenhouse*. Volume II. F. Warne & Co., London, New York.

Targioni Tozzetti, A. (1825). *Raccolta di Fiori Frutti ed Agrumi*. Florence.

Thomé, O. W. (1885). *Flora von Deutschland Österreich und der Schweiz*. Volume III. F. E. Köhler, Gera-Untermhaus.

Van Houtte, L. (1873). *Zea mays. Flore des serres et des jardin de l'Europe*. Volume 19, t. 922.

Yokohama Ueki Kabushiki Kaisha. (1907). *Catalogue of the Yokohama Nursery Co., Ltd*. Yokohama Nursery Co., Yokohama.

Art Collections

Marianne North (1830–90). Comprising over 800 oils on paper, showing plants in their natural settings, painted by North, who recorded the world's flora during travels from 1871 to 1885, with visits to 16 countries in 5 continents. The main collection is on display in the Marianne North Gallery at Kew Gardens, bequeathed by North and built according to her instructions, first opened in 1882.

FURTHER READING

Bynum, H. and Bynum, W. F. (2014). *Remarkable Plants That Shape our World*. Thames & Hudson, London in association with the Royal Botanic Gardens, Kew.

De Cleene, M. and M. C. Lejeune. (2003). *Compendium of Symbolic and Ritual Plants in Europe*. Man and Culture, Ghent.

Drori, J. (2020). *Around the World in 80 Plants*. Laurence King Publishing, London.

Giesecke, A. (2014). *The Mythology of Plants: Botanical Lore from Ancient Greece and Rome*. J. Paul Getty Museum, Los Angeles.

Henderson, H. (2005). *Holidays, Festivals, and Celebrations of the World Dictionary*, third edition. Omnigraphics Inc., Detroit.

Hutton, R. (1996). *The Stations of the Sun: A History of the Ritual Year in Britain*. Oxford University Press, Oxford.

Majupuria, T. C. and Joshi, D. P. (2009). *Religious and Useful Plants of Nepal and India*. Rohit Kumar, Kathmandu.

Mills, C. (2016). *The Botanical Treasury*. Welbeck Publishing, London in association with the Royal Botanic Gardens, Kew.

North, M. and Mills, C. (2018). *Marianne North: The Kew Collection*. Royal Botanic Gardens, Kew.

Payne, M. (2016). *Marianne North: A Very Intrepid Painter*, revised edition. Royal Botanic Gardens, Kew.

Roy, C. (2005). *Traditional Festivals: A Multicultural Encyclopedia*. ABC-CLIO, Santa Barbara.

Vickery, R. (2019). *Vickery's Folk Flora: An A–Z of the Folklore and Uses of British and Irish Plants*. Weidenfeld & Nicolson, London.

Watts, D. (2007). *Dictionary of Plant Lore*. Elsevier, Amsterdam.

ACKNOWLEDGEMENTS

Kew Publishing would like to thank the following for their help with this publication: in Kew's Library and Archives, Fiona Ainsworth, Craig Brough, Rosie Eddisford, Arved Kirschbaum and Anne Marshall; for digitisation work, Paul Little; and for their suggestions of plants and festivities, Henrik Balslev, Leigh-Anne Bullough, Caspar Chater, Pei Chu, Beatrice Ciută, Radu Costel, Marina Economou, Aisyah Faruk, Shahina Ghazanfar, Zoya Khan, Alasdair Nisbet, Avigail Ochert, Piya Rajendra, Sita Reddy, Leila Redpath, Jennifer Truelove, Tim Utteridge, Saskia Wolsak.

INDEX

First published in 2021
Royal Botanic Gardens, Kew,
Richmond, Surrey, TW9 3AB, UK
www.kew.org

ISBN 978 1 84246 725 1

Distributed on behalf of the Royal Botanic Gardens, Kew in North America by the University of Chicago Press, 1427 East 60th St, Chicago, IL 60637, USA.

British Library Cataloguing in Publication Data
A catalogue record for this book is available from the British Library

Design: Ocky Murray
Page layout and image work: Christine Beard
Production Manager: Jo Pillai
Copy-editing: Michelle Payne

Printed and bound in Italy by Printer Trento srl.

Front cover images: *Euphorbia pulcherrima* (see page 32), *Viscum album* (see page 88), *Zingiber officinale* (see page 52), *Phoenix dactylifera* (see page 40).

p2: *Ilex aquifolium*, holly, from Henri Louis Duhamel du Monceau *Traité des Arbres et Arbustes qui se Cultivent en France en Pleine Terre*, 1755.

p4: Flor de Pascua or Easter Flower at Morro Velho, Brazil, by Marianne North from the Marianne North Collection, Kew, 1872.

p8: *Olea*, olive, from Henri Louis Duhamel du Monceau *Traité des Arbres et Arbustes qui se Cultivent en France en Pleine Terre*, 1755.

p10–11: Moon Reflected in a Turtle Pool, Seychelles, by Marianne North from the Marianne North Collection, Kew, 1883.

For information or to purchase all Kew titles please visit shop.kew.org/kewbooksonline or email publishing@kew.org

Kew's mission is to understand and protect plants and fungi, for the wellbeing of people and the future of all life on Earth.

Kew receives approximately one third of its funding from Government through the Department for Environment, Food and Rural Affairs (Defra). All other funding needed to support Kew's vital work comes from members, foundations, donors and commercial activities, including book sales.

Publishers note about names
The scientific names of the plants featured in this book are current, Kew accepted names at the time of going to press. They may differ from those used in original-source publications. The common names given are those most often used in the English language, or sometimes vernacular names used for the plants in their native countries.

MIX
Paper from
responsible sources
FSC
www.fsc.org
FSC® C015829

Cacti

Palms

Honzō Zufu

Japanese Plants

Carnivorous Plants

Wildflowers

Fungi

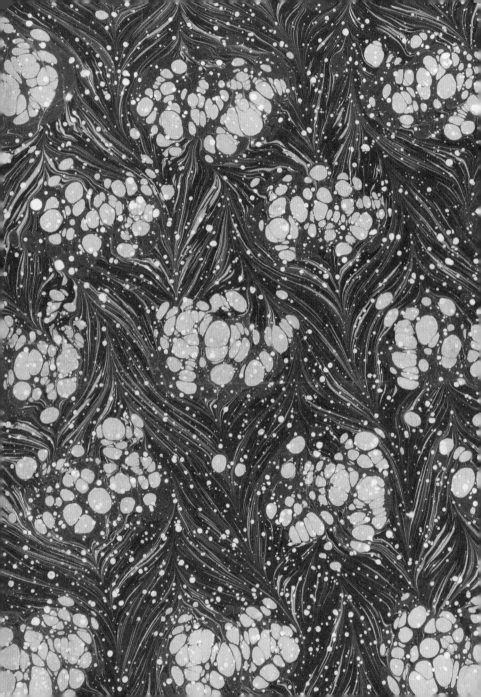